YOUR KNOWLEDGE HAS VALUE

- We will publish your bachelor's and
  master's thesis, essays and papers

- Your own eBook and book -
  sold worldwide in all relevant shops

- Earn money with each sale

Upload your text at www.GRIN.com
and publish for free

Suryansh Pant

SMART Grids: A Technology for Society

A short Introduction

GRIN Verlag

Bibliografische Information der Deutschen Nationalbibliothek:

Die Deutsche Bibliothek verzeichnet diese Publikation in der Deutschen National-
bibliografie; detaillierte bibliografische Daten sind im Internet über http://dnb.d-
nb.de/ abrufbar.

Dieses Werk sowie alle darin enthaltenen einzelnen Beiträge und Abbildungen
sind urheberrechtlich geschützt. Jede Verwertung, die nicht ausdrücklich vom
Urheberrechtsschutz zugelassen ist, bedarf der vorherigen Zustimmung des Verla-
ges. Das gilt insbesondere für Vervielfältigungen, Bearbeitungen, Übersetzungen,
Mikroverfilmungen, Auswertungen durch Datenbanken und für die Einspeicherung
und Verarbeitung in elektronische Systeme. Alle Rechte, auch die des auszugsweisen
Nachdrucks, der fotomechanischen Wiedergabe (einschließlich Mikrokopie) sowie
der Auswertung durch Datenbanken oder ähnliche Einrichtungen, vorbehalten.

Imprint:

Copyright © 2014 GRIN Verlag GmbH
Druck und Bindung: Books on Demand GmbH, Norderstedt Germany
ISBN: 978-3-656-82235-6

This book at GRIN:

http://www.grin.com/en/e-book/282436/smart-grids-a-technology-for-society

JACOBS UNIVERSITY

School Of Engineering and Science
International Logistics: Management & Engineering

Term Paper

SMART Grids: A Technology for Society

Suryansh Pant

USC: Society and Technology

May 17, 2014

Abstract

The following paper presents the 'Smart-Grid VDAR' technology presented at CeBIT, Hannover. The idea and structure of the technology is subsequently introduced followed by a social analysis of the technology. Some of the relevant groups identified were: consumers, producers, environmentalists and the government. High installation costs and health concerns were addressed as the main problems concerned with the use of technology. Finally, the aim of European Union to adopt a smart grid within the next decade is highlighted which is an indication of an overall positive outlook towards the technology.

Keywords: Smart-Grids, European Union, SCOT Analysis, Energy Market

Introduction

The governments of most countries started liberalizing their energy markets in the early 90's. This deregulated energy market gave customers the choice of selecting electricity providers according to their needs and prices. However, a fierce competition emerged amongst the providers to generate, transmit and distribute power.[1] Ignoring the characteristics and the scale of deregulation in different countries, two common patterns have been observed: 1) It is difficult to store energy and deliver on demand. 2) Demand is price insensitive. This means that the power demand remains the same or continues to go up even if the electricity prices increase largely.[2]

The two occurrences caused unreliability of power when providers and distributors were faced with a sudden increase in demand which ultimately meant high prices for customers.[3] To cope, many utility industries installed sensors and factories to analyse real-time energy consumption which enabled them to predict power demands up-to a certain point. As sensors became cheaper and wireless communication more widespread, the companies began avoiding power failures and overheating of power-lines through quickly rerouting electricity.[4] This is considered a 'smart-grid' whereby intelligent power supply networks permit efficient and reliable use of energy.

The following paper reports on the 'VDAR technology' which is an extension of basic 'smart-grids' on a national and international level. The information regarding the technology was encountered on a field trip to CeBIT, Hannover.[5] This paper analyses the technology by answering the following questions: -

    1)   What are Smart-Grids in VDAR IT Project?

---

[1] Cf. (Arentsen & Kfinneke, 1996)

[2] Cf. (Ding et al., 2013)

[3] Cf. (Wang et al., 2011)

[4] Cf. (Scientific American, 2011)

[5] CeBIT is the world's largest and most international computer expo. The trade fair is held each year on the Hannover fairground, the world's largest fairground, in Hanover, Lower Saxony, Germany, and is considered a barometer of the state of the art in information technology. It is organized by Deutsche Messe AG. The passes to the fair were generously provided by Volkswagen AG.

2) What is the social construction of Smart-Grids?

3) What is the future of Smart-Grids?

A thorough literature review is the basis of this report where papers written by presenters of the technology in Hannover, Germany remain the main focus.

What are Smart-Grids in VDAR IT Project?

The current power market is mainly an economic system which reflects the power flow in terms of trade for the day ahead and real time auctions. The VDAR IT Project proposes a 'smart-grid' ready power market which integrates the physical reality of the power market into the economic reality of the market model. This new grid is envisaged to contain a virtual grid based on bidirectional communication processes including novel components such as storage and distributed energy resources[6]. (see Figure 1)

Figure 1: The structure of Smart-Grids

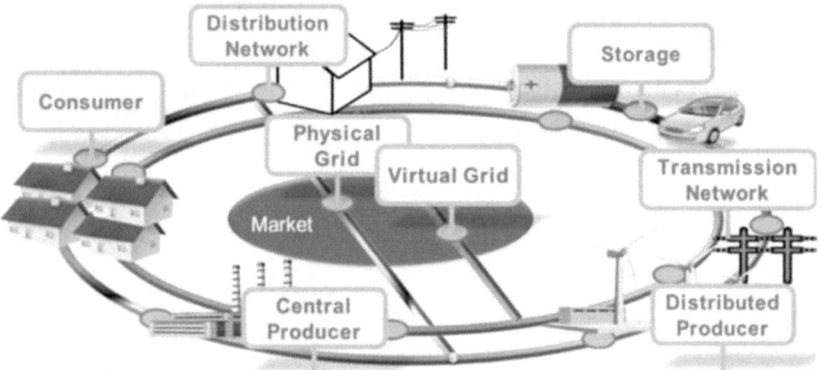

Source: VDAR IT Project, Karlsruhe Institute of Technology, CeBIT Hannover

The physical grid, the virtual grid and the economic market are interconnected by a decoupled control loop. In this 'smart-power-grid', consumption control is not about energy saving but about efficiently balancing supply and demand, ideally in real time, to prevent extreme load situations both in the market and the grid. For example, in Germany, this technology could be incorporated in a top-down approach in various states and region and

---

[6] Cf. (TECO, Karlsruhe Institute of Technology, 2012)

equally to cities, offices and home. A sudden demand leading to high local load and extreme prices can be compensated by borrowing and transferring power from other regions. Thus, in future smart cities, where anyone can be both a consumer and a producer, everyone can potentially contribute to the alleviation of price insensitive high demand across the network.

What is the Social Construction of Smart-Grids?

The Social Construction of Technology states that it is the human needs and actions which shape technologies and not vice-versa.[7] The SCOT (Social Construction of Technology) theory has 3 core principles namely: *'Interpretative Flexibility'*, which consists of how people perceive a technology, *'Relevant social groups'*, people associated with the technology and finally, *'Problems and Conflicts'*, which can arise. Smart-Grids are analysed on all three basis and it is hypothesized that the technology would have wide ranging influences on many social groups. The structure of the current power market is a good place to identify said social groups. (see Figure 2)

Figure 2: Actors in a Power Market

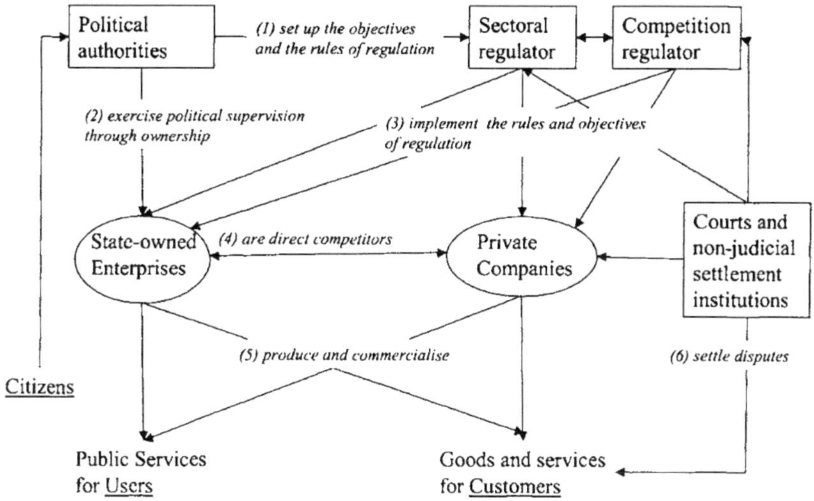

Source: (adapted from Arentensen, Kfinneke, 1996)

---

[7] Cf. (Bijker et al., 1993)

1   Interpretative Flexibility

The two main social classes which will be affected by 'smart-grids' are consumers and producers. The consumers of Smart grids can be broadly categorized into individuals, corporations and city/state. The consumers are most likely to view smart-grids favourably since they will have decreased costs from electricity bills when the power demand is high. Further, through the introduction of 'smart-meter' which relay information to power station, the consumers will be able to run home/office appliances which will work when the demand is low.[8]

On the other hand, the producers and distributors of electricity are more likely to view the technology unfavourably as it would cut into profits. Currently, the producers and distributors charge high prices when demand exceeds supply. With the advent of smart-grids, power could be borrowed from other regions which would keep costs low.[9]

2   Relevant Social Groups

Two relevant social groups identified were 'Environmentalists' and the 'Government'. The Environmentalists are confirmed to support the integrated smart grid technology as it consumes energy efficiently. It is a reliable source of energy which *"...can reduce carbon dioxide emissions by between 11 pounds and 110 pounds per customer per year."*[10]

Further, the government is likely to support and fund the activity as it will reduce pressure on traditional sources of energy such as fossil fuels. For example, the smart energy project in Germany received approximately € 140 million to install and test the technology of smart grids. Currently, six pilot projects are being supported by the Federal Economics Ministry in collaboration with the Federal Environmental Ministry. Germany has also launched research into a transnational smart-grid across the European Union which could potentially save the energy crisis of the future.[11]

---

[8] Cf. (IEE Smart Grids, 2013)

[9] Cf. (NAED, 2011)

[10] Cf. (Smart Grid Consumer Collaborative, 2013)

[11] Cf. (Smart Energy made in Germany, 2012)

3 Problems and Conflicts

Concerns have been raised regarding the use of 'smart-meters' in households across U.S. It is estimated that an average household has 10-15 appliances which operate at any given time during the day or night. According to recent studies, a smart meter operates at frequencies between: 917 MHz and 3.65 GHz which are well above the safety standards determined by health laws. Compounded with the frequencies from appliances of neighbouring household, this 24/7 RF exposure could potentially cause cancer and has raised health concerns amongst users.[12]

An additional cause of concern is the high costs of installation which comes with the introduction of smart grids. It is largely due to the fact that individual co-generators have higher costs than one large central generator. *"Deployment of smart grid technology from U.S. utility control centres and power networks to consumers' homes could cost between $338 billion and $476 billion over the next 20 years"*[13]

What is the Future of Smart Grids?

"Using thousands of miles of high-tech undersea cables the 'Super-Grid' will unite wind farms on blustery British coasts with Dutch and Belgian tidal power, the vast hydroelectric potential of Norway fjords and Germany's massive solar arrays."[14] The eight big countries of the EU: Germany, France, UK, the Netherlands, Luxembourg, Denmark, Sweden and Ireland are committed to having a working Super-Grid within the next decade. While the project sounds ambitious, if successful, it would fulfil the EU pledge to source 20% of its electricity from renewable sources of energy by 2020.

Further, smart grids are touted to the promotion of electrical vehicles which would in turn be a boost to the economic industry by creating more jobs. For example, U.S. Department of Energy estimated that if the Electrical Vehicles were adopted it would reduce foreign oil

---

[12] Cf. (Counter Punch, March 2011)

[13] Cf. (Energy and Environment, New York Times, 2011)

[14] Cf. (Eco Solutions, CNN, 2010)

imports by 52%. To achieve this goal, smart-grids are an absolute necessity as they could potentially accommodate millions of Electrical vehicles on the road.[15]

Lastly, we can safely say that as we become more and more globalized and environmentally conscious, it is our need for a greener environment and collective conscience which has shaped the technology of smart grids. Thus, the social construction of technology is inadvertently proven to be correct.

---

[15] Cf. (US DOE, 2013)

References

Bergman, L., Brunekreeft, G., Doyle, C., von der Fehr, N.-H., Newbery, D.M., Pollit, M. and R6gibeau,   P. (1998). *Europe's Network Industries: Conflicting Priorities, Monitoring European Deregulation 1: Telecommunications*. CEPR, London.

Arentsen, M. and Kfinneke, R. (1996). *Economic organization and liberalization of the electricity industry*. Energy Policy 24(6): 541-552.

Yong Ding, Per Goncalves Da Silva, Martin Alexander Neumann, Lin Zhang, Michael Beigl, *A Control* (B. Blake Levitt, 2011) (Behr, 2011) (Ford, 2010) (NETL, 2013)*Loop Approach for Integrating The Future Decentralized Power Markets and Grids*, 4th International Conference on Smart Grid Communicatio (Durand, 2013)ns (SmartGridComm), IEEE, October 2013.

T. Jamasb and M.Pollitt, *"Electricity Market Reform in the European Union. Review of progress towards liberalization & Integration,"* Energy Journal, vol. 26, no. SPEC. ISS., pp. 11-41, 2005.

G. Wang, A. Kowli, M. Negrete-Pincetic, E. Shafieepoorfard, and S. Meyn, *"A control theorist's perspective on dynamic competitive equilibria in electricity markets,"* in Proc. 18th World Congress of the International Federation of Automatic Control (IFAC), Milano, Italy, 2011.

The Smart Grid Opportunity . (2011). Retrieved May 5, 2014, from National Association of
   Electrical Distributors: http://www.naed.org/smartgrid/

NETL. (2013). Retrieved May 7, 2014, from U.S. Department of Energy:
   http://www.netl.doe.gov/research/energy-efficiency/energy-delivery/smart-grid

American, S. (2011, March 30). *What is the Smart Grid?* . Retrieved April 29, 2014, from
   YouTube: https://www.youtube.com/watch?v=-8cM4WfZ_Wg

B. Blake Levitt, C. G. (2011, March). The Problems with Smart Grids. *Counter Punch*, pp. 18-
   20.

Behr, P. (2011, May 25). New York Times. *Smart Grid Costs are Massive, But Benefits will Be Larger, Industry Study Says*.

Durand, P. (2013). *Smart Grid Economic and Environmental Benefits*. USA: Smart Grid Consumer Collaborative.

Ford, M. (2010, January 31). CNN, Eco Solutions. *Getting Connected: Europe's Green Energy Super Grid*.

IEE. (2013, September 4). *Smart Grid Consumer Benefits*. Retrieved May 3, 2014, from IEE Smart Grid: http://smartgrid.ieee.org/questions-and-answers/964-smart-grid-consumer-benefits

Roesler, D. P. (2012). *Smart Energy Made in Germany*. Berlin: B.A.U.M Consult GmbH, Munich/Berlin.

TECO. (2012, November 5). *VDAR-IT Project*. Retrieved May 1, 2014, from Karlsruhe Institute of Technology: http://www.teco.edu/research/vdar/

Wiebe E. Bijker, T. P. (1993). Common Themes in Sociological and Hostorical Studies of Technology. In *The Social Construction of Technological Systems* (pp. 9-83). London: The MIT Press .